A SCIENTIST SPEAKS

EXCERPTS FROM ADDRESSES BY

KARL TAYLOR COMPTON

DURING THE YEARS 1930-1949
WHEN HE WAS PRESIDENT OF
THE MASSACHUSETTS INSTITUTE OF TECHNOLOGY

PUBLISHED BY THE UNDERGRADUATE ASSOCIATION
MASSACHUSETTS INSTITUTE OF TECHNOLOGY
CAMBRIDGE
1955

Acknowledgment
is here made to Miss Frances Van Schaick
for the taste and judgment with which she
assembled the text of this book.

FOREWORD

The Undergraduate Association of the Massachusetts Institute of Technology has published this book not as a memorial to the administrator who guided the development of M.I.T. from 1930-1954, nor to the physicist who made lasting contributions to scientific advancement, nor to the patriot who personally shortened the duration of World War II, but as a remembrance to a friend, with the hope that the thoughts expressed on the following pages will help to keep alive in the hearts of those who knew him the spirit and the ideals that are Karl Taylor Compton.

Eldon H. Reiley
President of the Undergraduate Association

January 12, 1955

PREFACE

This volume of excerpts from addresses of
Karl Taylor Compton represents the fulfillment of a dream
long cherished, of preserving for the record the expression
of ideas which, deeply felt as he spoke them, stand today
as witness and memorial of his greatness as man, scientist-
teacher, and statesman.

The editorial decisions which gave form
to the present volume involved choices not always easy
to make. In adopting this form of presentation we have
taken responsibility for extracting excerpts from addresses
over a twenty-year period, and establishing them in new
sequences. Every effort has been made to keep them in a
context which preserves the spirit of their original
presentation.

By not dating the individual excerpts, we
have, perhaps, sacrificed justice to art. For most of the
ideas and recommendations which are here included, were
originally expressed over twenty years ago and gain significance from that fact. The years covered by these speeches
saw unprecedented changes in the world atmosphere and in
popular attitudes. The period from the midst of the depres-
sion through World War II to the mid-century point en-
compassed changes that in other eras would have taken
centuries. The volume remains to be written which will
tell the story of Karl Compton the influential citizen who,

through his various achievements as scientist, educator, and statesman helped to bring many of his pioneering ideas of yesterday into the realities of today.

Nor shall we attempt to tell this story in this introduction, except through one quotation which is suggestive of the whole man and his works. In a speech telling of how the invitation came to him in 1930 to be President of M.I.T., Dr. Compton recalled:

"I was not too sure I wanted to be president of M.I.T. . . . . I wanted to think it over and to talk it over with Frank B. Jewett, President of Bell Laboratories . . . and . . . made an appointment with him. As I left the house that morning I told my wife that I was pretty sure I would turn the job down; and in fact, Dr. Jewett did nothing positive to try to influence me. He said that he had observed some things about engineering schools in the United States. They had performed a useful function several decades ago and had done a marvelous job, in his opinion, but that they were far behind in their usefulness in the present and particularly in any future. They were too much on the pattern of technical trade schools. The only solution was to try to inject into the policies of these institutions a greater interest in fundamental science and research, if someone could be found to do it. So I thought it was up to me to try to help."

Karl Compton was, and the memory of him is, inspiration and stimulation to all of us at the Institute. This little volume is an expression of the immeasurable debt we owe him.

J. R. Killian, Jr.

January 12, 1955

CONTENTS

I THE AGE OF SCIENCE

We live in an age of science. I do not say
"an age of technology" for every age has been an age of
technology. We recognize this when we describe past civ-
ilizations as the Stone Age, the Bronze Age, and the Age
of Steam or of Steel, thus implicitly admitting that the
stage of civilization is determined by the tools at man's
disposal—in other words, by his technology. . . . Science,
unlike invention and technical skill, is a relatively modern'
concept.

The characteristic feature of our age results
from the wedding of science and engineering. It is the working together of disciplined curiosity and purposeful ingenuity to create new materials, new forces, and new opportunities which powerfully affect our manner of living and ways
of thinking.

Neither curiosity nor ingenuity is a modern
impulse. . . . The distinctive feature of science and technology at the present time is the accelerated pace of their
development. This is partly due to continually improved
techniques and organization, and it is partly due to the
great accumulation of knowledge and art, because the

more information and tools we have at our disposal, the more powerful can be the attack on any new problem.

Fundamentally, science means simply knowledge of our environment. Combined with ingenuity, science becomes power.

I believe that the advent of modern science is the most important social event in all history.

The geographical pioneer is now supplanted by the scientific pioneer. . . . Without the scientific pioneer our civilization would stand still and our spirit would stagnate; with him mankind will continue to work toward his higher destiny. This being so, our problem is to make science as effective an element as possible in our American program for social progress.

There is "something new under the sun" in that modern science has given mankind, for the first time in the history of the human race, a way of securing a more abundant life which does not simply consist in taking it away from someone else. Science really creates wealth and opportunity where they did not exist before.

Whereas the old order was based on competition, the new order of science makes possible, for the first time, a co-operative creative effort in which everyone is the gainer, and no one the loser.

It is hard, now-a-days, to realize how revolutionary have been the changes in man's outlook upon life which have been wrought by science.

With the advent of science came the beginnings of a profound change of intellectual approach to the interpretation of Nature. Supernatural causes, manufactured in the imagination to satisfy the desire for explanations, were superseded by natural causes discovered in the laboratory. The universe came to be understood as a vastly complicated mechanism built of electricity and energy, and operating with marvelous precision in accordance with a system of laws.

As the complexity of the structure of matter became revealed through research, its basic simplicity, unity, and dependability became equally evident. So we now see ourselves in a world governed by natural laws instead of by capricious deities and devils. This does not necessarily mean that God has been ruled out of the picture, but it does mean that the architect and engineer of the universe is a far different type of being from the gods assumed by the ancients, and that man lives and dies in a world of logical system and orderly performance.

This change in our understanding of the world has not only profoundly affected our conceptions of the place of man in the universe and his attitudes toward it, but it has also exerted an influence on his political organization.

I think we can certainly say that, insofar as development in science is concerned, we are now in an era of vigorous development both of science itself and in the techniques of organization, administration, and support which form the environment in which scientific progress can be made.

II OPPORTUNITY AND RESPONSIBILITY

I would emphasize the fact that scientific discovery is, *per se*, neither good nor bad. It simply produces knowledge and with knowledge, opportunity and responsibility. I think it fair to say that the advance of science carries with it powerful demands on morality if the results are to be beneficial rather than harmful.

How true it is, as President Bowman of Johns Hopkins once said, that "every man who puts fire and light into the hands of men puts also beside them a wrath and a violence, a gift and a penalty." For in a double sense the fire and the light of science—its powers that have made a new world, its intellectual triumphs that have illuminated that world with the rays of reason—have also made more hazardous the ordeal of living in a day when wrath and violence would seize for their destructive purposes the very tools and services brought down from the skies as the means of a fresh creation.

Because of scientific discovery, society requires as never before a psychology of faith and courage, a spirit of unselfishness and co-operation, a high degree of intelligence among citizens and leaders alike, and an ethics far closer to the principles of Christ than our so-called Christian nations have in the past practiced.

I like to think of culture as signifying a sympathetic understanding of life, and in this sense it is evident that the cultured person of the future will be one who includes among his interests and attainments a basic knowledge of science, a deep respect for the rights and privileges of the less fortunate, a desire to help those who need and deserve it, encouragement for those on the threshold of tomorrow's successes so that they may look forward to a healthier existence through science, through religion, through civic response, through mutual respect and understanding.

Attempts to discredit methods and results of science are continually being made on the part of certain doctors of society whose motive power is a blend of about equal parts of laudable desire to reform society, of nervousness over existing conditions, and of ignorance regarding the real accomplishments and possibilities of science.

That man who is afraid of the opportunities thus made available because of the difficulties which he sees in the way, is to be likened to the man who received from his master one talent, and being afraid of the responsibility, wrapped it in a napkin and buried it.

The great significance to me of the parable of the talents is that the man who buried his talent was afraid to take the responsibility of using his opportunity. To my mind, applied science is like the ten talents which have been given us to use as profitably as we can. It is an opportunity of tremendous possibility for the good of humanity.

It is up to each of us, singly and collectively, to see to it that the result is the great good and not the great harm.

I have become continually more and more impressed with the direct benefits and opportunities in science for increased effectiveness in our ways of living and working, and with indirect benefits which should accompany this in the directions of greater time and opportunities for cultural development if we but have the moral urge to use the opportunities aright.

Does this technological age bring greater human happiness than, for example, the Periclean Age of the ancient Greeks? Now man's enjoyment of Periclean Greece would depend on his political status, on whether he was a member of the small aristocratic class that had leisure and freedom for thought based upon the productive power of thousands of human slaves, or whether he was one of those slaves. But in this day and generation, machinery is the slave of all men.

To an unprecedented extent all men may have opportunities for education, recreation, and the nobler

7

things of life because machinery is doing most of the drudgery for them. If man had to supply the energy that America uses from waterpower, coal, and gasoline, we should need the exhausting labor of five thousand million slaves! Such is the debt of modern man to technology.

The greatest factor for spiritual and ethical improvement is education. But it is only because of increased productive power due to science that it is possible to devote the first twenty years of life to education, instead of to the struggle for existence. Similarly it is science which has lengthened the average span of human life from thirty-three to sixty years during the past 140 years. Think what this has meant in alleviation of suffering and in the rounding out and the satisfaction of living!

Scientific discovery is a search for knowledge, one of man's noblest occupations. And the applications of science are, in large proportion, in the interests of human welfare. As my friend Dr. Hu Shih once said:

To me that civilization is materialistic which is limited by matter and incapable of transcending it; which feels itself powerless against its material environment, and fails to make the full use of human intelligence for the conquest of nature and for the improvement of the conditions of man. . . .

On the other hand, that civilization which makes the fullest possible use of human ingenuity and intelligence in search of truth in order to control nature and transform matter for the service of mankind, to liberate the human spirit from ignorance, superstition and slavery to the forces of nature, and to reform social and political institutions for the benefit of the greatest number—such a civilization is highly idealistic and spiritual.

Applied science is not an end in itself, but it is the most powerful means ever discovered for supplying the opportunity to secure the finest things of life.

Why not speak of this as the Age of Man's Opportunity?

III ENVIRONMENT FOR SCIENCE

Those of us who are greatly interested in the contribution which science can make to national welfare must have also an intelligent and effective interest in those environmental aspects which not only affect the opportunities for scientific progress, but affect every aspect of our national life and individual happiness.

Scientific work and scientific workers and scientific institutions cannot survive and prosper unless the general environment is favorable. . . . Science does not live to itself; it lives only as it contributes usefully to other aspects of society, and it lives only if certain other essential aspects of society are in a healthy state to provide the necessary environment.

These other aspects . . . involve such matters as international peace, the healing of the wounds created by the war, the relations between management and labor which if handled successfully, can lead to a most productive team-work for the benefit of all and which, if handled unsuccessfully can lead to national, or even international chaos. They include governmental policies, such as the extent to which Mr. John Q. Citizen is to be regimented, or to which he is to retain a high degree of freedom of opportunity and initiative. They include the financial policies which can bankrupt the nation, or through inflation, lead to the wiping out of all the reservoirs for free enterprise, if unwisely handled.

The whole history of scientific progress illustrates the importance of free communication of ideas, of co-operative work at all levels, of adequate support and facilities, and above all, of high grade research workers and top-notch leadership.

It is especially important at the present time for us to understand the international character of science, sometimes expressed by the phrase, "Science knows no national boundaries." Of the many reasons for this I shall mention only two: we must understand the conditions under which science can flourish in America if America is to benefit through science to the maximum extent in industry, health, and security; we must quell the idea which seems to be forming in some quarters that, because science is international, scientists are not to be trusted in matters of loyalty or patriotism.

When I was a young instructor in a western college, its president urged me to offer a special course in physics for the women students. I stoutly maintained that the facts of nature are the same for women as they are for men. Similarly, the facts of nature, whose discovery is the objective of science, are the same for the

American as they are for the Italian or the Russian or the Chinese; furthermore, the methods which are successful in discovering these facts of nature in one country are equally good in another. In this fundamental sense science knows no national bounds.

While we are all greatly concerned with our national security in this troubled post-war world, it is important for us to think clearly on the subject of national security and what can provide it. So far as science and its applications are concerned, *national security is by no means synonymous with secrecy*. In fact, the rigid imposition of a high degree of secrecy on American scientific progress in any field, even atomic energy and nuclear science, would be about the most disastrous policy that our country could pursue from the national security standpoint. The reason, of course, is that progress cannot flourish under conditions of secrecy or even suspicion.

The important point of view to get is that *national security is achieved by our being as far ahead as possible of any unfriendly competitor*. To be far ahead, the methods by which science can make progress must be followed, and these are not the methods of secrecy. I put this strongly even though I am fully aware that there are a number of important facts which should be very carefully safeguarded by security methods and by severely limiting the number of persons having access to these facts. However, these are exceptional cases. Our natural tendency in official quarters will be to err on the side of too great secrecy and supervision rather than too little.

It would be better to take the carefully calculated risk of allowing some confidential information to get out of our hands if such a policy would enable us to advance our own science and art at a rate which a competitor cannot hope to equal than it would be to impose by regulation or public opinion a condition which seriously handicaps progress by rendering employment in these pursuits definitely unattractive to top-flight scientists and engineers who have plenty of other opportunities to turn their talents into more comfortable and usually more rewarding directions.

I am one of the large group of scientists and engineers who are strong advocates of national military preparedness at this time, but who are nevertheless worried over the detrimental effects on our work for technological preparedness of the . . . charges which have been made against reputable scientists on the basis of hearsay and unsubstantial evidence.

There are a number of things which the United States must do if it is to guard against the danger of future war. To be prepared in a military sense is only one of these things. Even more important, the United States must be just, reasonable, and co-operative in its dealings with other countries. And in order that the burden of preventing wars may not fall too heavily on any one country, it is important that all countries which are determined to have peace in the future should combine to insure it. But even in such a combine or league, the effort will prob-

13

ably be fruitless unless there are firm intent and adequate strength to enforce good behavior on any nation which becomes obstreperous in a warlike way. So I believe that a reasonable degree of military strength is one of the essentials to future peace.

A citizens' army has always been preferred in the United States to a large professional army. Technological developments have greatly changed the conception of an effective citizens' army. For one thing, war on land, sea, and air has become so highly technical that a much longer period of training is necessary than was true fifty or one hundred years ago. The training itself must be largely technical. More important still, the speed of transportation and the development of methods for making powerful attacks with great suddenness and at great distance mean that it is no longer safe to wait until war breaks out to begin the intensive training of our armies. These factors, I take it, have all been influential in leading the military experts to the conclusion that a year of military training is needed to provide a continuously replenished reservoir of men who could be quickly called to arms for effective military service at threat of war.

To be a stable and effective citizen requires a nice balance between freedom and initiative on the one hand, and discipline on the other. In this modern age we scarcely need to make an argument for freedom and initiative because certainly the youth of our country, by and large, have a great deal of both. The serious question, in

14

fact, is whether they do not have too much freedom and too little discipline to be the most effective possible members of society.

It is only the man who lives entirely alone who has complete personal freedom. Every social contact requires some adjustment, and every effective adjustment is a discipline. I believe that one year in the stage between youth and manhood spent in the universal military training program would pay good dividends in this matter of discipline. It would develop fundamental habits of promptness and precision, it would develop the ability to work as part of an organization, it would develop the attitude of teamwork which I think is one of the best examples of what is meant by real discipline.

Discipline in the sense in which I am using it does not mean subjection of the individual. It means smooth and automatic action by the individual in his routine situation so that he can be free to give attention to the new and unusual problems which may face him. It is this type of discipline which is fundamental in an army, and it is this element of training and character which I believe a year of universal military training would strengthen in our oncoming generation of citizens.

What is most needed is to recapture in the public mind and in the attitudes of the scientists who must do the jobs at hand, that spirit of confidence in, and enthusiasm for work for their country, with which we emerged from the war. . . . If the recent trend of suspicion and lowering morale continues, we shall be in really serious danger.

I should like to emphasize the responsibilities of society for handling the opportunities created by science, rather than the responsibilities of science to society. For it is science which provides the opportunities, but it is society which must take the responsibility of handling them.

IV NEW PERSPECTIVES

The alteration in perspective which has come about as a result of achievements in . . . science, bringing gradually into focus our picture of the world and the universe, and of our own place therein, is a most salutary one, salutary both as it clips the wings of foolish fancy, and as it spreads to full sweep the far more powerful wings of sober fact.

The fact that the human intelligence is ready to range so widely as from grappling with the microscopically invisible virus to measuring the unimaginable interstellar space, is in itself cause, not for silly pride, but for honest admission of achievement.

Objective reason is the greatest safeguard against mere vaunting egotism, yet it is also the surest way to recognition of simple worth. If we can occasionally stand at a safe distance and look at the odd and defenseless beings which we are, and observe the will which has enabled such puny things to do so much, we are forced to this recognition.

It is perhaps among the greatest of paradoxes that the extension of knowledge which has shown man to be, not the center of creation occupying an earth which is the center of the universe, but rather only one more form of life on a planet comparatively infinitesimal, should at the same time increase, justly and honestly, his sense of his own dedication to something beyond mere being, and his faith that life as manifested in him possesses worth and significance beyond the measure of the very knowledge which seems at first thought to reduce life to formulae. Yet so it is.

But little reflection is necessary to enable us to see why this seeming paradox is no paradox at all. As science has pushed the boundaries of knowledge farther and farther back, and has brought more and more out of the shadow of speculation into the light of objective fact, it has shown life to be infinitely more complex, infinitely more mysterious, far more inextricably involved in environmental, physical, physiological, psychic, and psychological issues. It is no more than natural for the logically trained mind, reasoning coldly by inference and association, to conclude that the spiritual, the non-material aspects of life may well also be far more ramified, far more mysterious, than would at first appear.

If humility is the mark of faith, then science has augmented the dignity of life by giving faith greater reason than ever for humility.

Science has not supplanted or destroyed religion, and cannot. It can, however, give a setting to which our thoughts on religious matters must conform. In its history, science has continually forced men to take an ever wider and grander concept of religion, by breaking down artificial barriers of ignorance and superstition. Its whole tendency has been to emphasize the fundamentally spiritual character of religion. . . . Science has had tremendous influence in shifting the field of religion from the physical to the spiritual world. We must not shut our eyes to the possibility of still further influence.

Science has contributed to the making of religion into a developing, dynamic spiritual force.

It is, perhaps, one of the great indirect results of modern science that there is gradually developing more of a spirit of religious tolerance, in which, outside of certain fundamental religious concepts or attitudes, the details of creed and doctrine are considered more in terms of individual preference than as boundary lines between damnation and salvation. We recognize the fact of wide differences of temperament and mentality between people, differences so great that no one system of religious experience or motivation or belief could possibly be a powerful influence for all.

I would make it clear that I believe religion to be intensely personal: no man can undertake to specify the religious ideas of another. In fact that is the basic principle of "freedom of worship." . . . In many ways a man's religion is a function of his own environment, his emotional and rational make-up. Some aspects I believe are fundamental and universal; but many aspects of religious form, expression, and impact depend on the personality, background, and environment of the individual. I want to make it clear that I recognize these differences, that I believe they are inevitable and valuable, and that I would not undertake to prescribe for any one his beliefs or his attitude toward religion in general or any religious creed or doctrine in particular.

There seem to be four great motives for religion:

(1) Desire for development and expression of a moral code as a basis for satisfactory human relationships.

(2) Longing for a source of spiritual strength, especially when man feels his own inadequacy. This may include desire to escape a feeling of guilt.

(3) Desire for immortality.

(4) Desire to explain the world of nature. These are all quite independent of each other, but they have all become embodied in almost every religious creed. To my notion only the first two have any proper role in religion.

The . . . explanation of the world of nature belongs not to religion but to science, and the sooner this is fully accepted the better it will be for both. . . . The . . . desire for immortality seems to me to be a subject for speculative philosophy, not for religion.

20

Whatever may be the scientific facts of immortality, still unknown, one thing is certain: every person is immortal in the sense that his influence continues forever,—the influence of his achievements, the effects of his ideas and attitudes, and the continuance of his flesh and blood as well as his spirit through his children. Truly, "no man liveth to himself alone."

Let me mention still another divergence of religious thinking, sometimes called fundamentalist and liberal. I prefer to think of them as the static and the dynamic concepts of religion. The impact of science on religion has served to emphasize the dynamic as opposed to the static concept. An extreme example of static attitude is implicit belief in the literal accuracy and permanent perfection of the Scriptures of the Jews or the Bible of the Christians. The dynamic attitude is to view those documents as the story of man's continual progress in evolving a religious attitude toward his environment and all that this environment implies. With this viewpoint, grotesque contradictions disappear. We see the evolution of his idea of God from an anthropomorphic conception of numerous deities of capricious behavior and often conflicting purposes, through the notion of a single God who walked and bargained with men, who chastised them and repented, to the conception of a great spiritual force operating through natural laws which are understandable and dependable and at least partly discoverable through science. We see the picture of a continual development in ideas of right and wrong from the early notions of obedience to sets of rules to concepts of social justice and human welfare. We

see notions of salvation and eternal life becoming less con-
centrated on selfish considerations and more concerned
with service to others and the permanent contribution of
our individual lives to the future welfare of mankind. This
dynamic concept of religion as a continually evolving and
developing spiritual force is inspiring and acceptable in a
scientific world. In my judgment the static concept of religion is sterile, discouraging, and unacceptable.

I think that we can agree on certain basic
facts: The religious impulse is instinctive, and a fundamental part of man's psychological, emotional, and mental
make-up. It includes the natural instinct toward a higher
moral code and it tends to devise methods for implementing
this instinct so as to achieve the desired results. It includes
the longing for spiritual strength and satisfaction, which
finds expression in such attitudes as reverence, worship,
and prayer,—though the worship and prayer may vary
greatly among different people and sects in degree of for-
mality and type of expression. It has always been, and I
believe always will be the most powerful influence for good.
The great permanence of religious ideas and sects inevitably
carries with them a great deal of tradition, some of which
may become obsolete. But all religions worthy of the name
possess common basic ideals of goodness, of unselfishness and
service, of reverence for a power which transcends our
human strength and understanding. A man can be a religious man without necessarily accepting all the beliefs
or dogmas of any given church or sect. But many people
find comfort and help in various beliefs, forms, or rituals

22

which to other people may not seem valid or satisfactory. In each case, I believe, the form and expression of man's proper religious life are those which give him the greatest spiritual comfort and inspiration toward the better life.

The gradual development—which has come quite rapidly in our own times—of the social sciences as a set of disciplines combining the traditional ethical and spiritual outlook of the humanities with the accurate and objective method of the physical sciences, is of profound import. From this development, even in spite of the distress and near disaster confronting us today, may be expected amelioration of abstract conditions of life fully as great as the improvement of the concrete conditions of living which the past three or four generations have witnessed.

The ultimate recognition of the truth that man is himself a part of the physical world, and subject to the laws which appear to govern it—the recognition which a fully self-conscious social science must bring in its train—cannot but lead to pronounced changes in many of our ingrained beliefs, and in many of our accustomed institutions.

The greatest of all contributions of science is found not in the comforts, pleasures, or profits which flow from it, but in the freedom and imagination which it has brought to the human spirit and the sense of relationship and unity in the world.

23

V ANALOGIES FROM SCIENCE

One of the fundamental laws of physical science is that Nature, if left to herself, moves in the direction of chaos. This is a somewhat crude way of stating the second law of thermodynamics. If the molecules of air in a room, for example, were all given uniform motions, and then left to themselves, their interaction with themselves and the walls would quickly change their velocities into the most random possible arrangement of motions, sometimes called the "Maxwell distribution of velocities."

In the field of human affairs many analogies may be drawn with this second law of thermodynamics. Human affairs, if left to themselves, also tend toward chaos. A business organization, if left without a guiding hand, becomes a disorganized business. A farm, if left to itself, becomes a wilderness. An economic policy of "let Nature take her course" leads inevitably to economic chaos. A political policy of "laissez faire" can lead nowhere but towards anarchy.

It is a significant fact that, in physical science, only one way has ever been suggested by which the tendency toward chaos can be circumvented. . . . This way is by the exercise of intelligence in carrying out a planned policy. It was Maxwell who showed how this might be done in the case of molecules of a gas through the agency of a hypothetical intelligent being, who has been dubbed "Maxwell's Demon," and who operates to sort out the fast molecules from the slow ones by means of a vessel provided with a trap door, which he opens to each fast molecule and closes against each slow molecule as it approaches. Thus contrary to the second law of thermodynamics, he substitutes a systematic for a chaotic distribution of molecules and thereby creates a store of energy because the fast molecules inside the vessel are at a higher temperature than the slower ones on the outside.

The analogy here with human affairs is apt. It is only by intelligence, creating and carrying out a plan, that a business, a farm or a government can be made productive instead of degenerating into disorganization and impotence.

Without long-range planning, society is not like a man but like an animal—impulsive, selfish, greedy, without goal or purpose—lovable, perhaps, but relatively ineffective.

"Letting Nature take her course" was listed above as one of the paths to chaos. Yet there is a very fundamental truth in the minds of some people who express their economic philosophy in this phrase, namely, that experi-

25

menting or manoeuvering, to be successful, must be directed to take advantage of natural laws and not to run counter to them.

The achievements of science come by the intelligent, skillful utilization of established laws of science. Similarly, experiments in human affairs will be successful if they are planned in accordance with, and not contrary to, basic social laws. If, as is unfortunately so often the case, we do not yet know what these laws are, then the "trial and error" method is the only recourse. The success of this method depends on a combination of knowledge, judgment, skill, and luck.

I can only venture to suggest some general principles of government which seem to me to be fundamental in utilizing and preserving the scientific temper. The first of these is that a degree of centralized government is an essential attribute of decent group life. Without it there is chaos, discord, and ineffectiveness. . . . Undoubtedly the increasing complexities of modern life due largely to technological progress require a continually increasing degree and quality of group control.

The second principle is, at first glance, the antithesis of the first. It is that wise government involves the minimum of control and supervision consistent with reasonably smooth, co-ordinated, and properly oriented operation.

26

My experience as a scientist has amply demonstrated that . . . the most significant technological ad-
vances have not come out of direct efforts to make them,
however well organized, but as unexpected by-products of
scientific work undertaken for quite other objectives—
usually for the satisfaction of scientific curiosity. Organized,
directed effort is very effective in perfecting the details of
a product or its production, but not in its initial discovery.
This contrast is greater the more epoch-making is the new
discovery.

Really epoch-making discoveries are relative-
ly unpredictable in advance. . . . Really new ideas do not
come to order, and are not pulled out of a hat, and who
can tell in whose brain they will germinate? If an industrial
research laboratory had been established a century ago to
improve lamps, it would have investigated inflammable oils
and gases, wicks, chimneys, and refractories. Not conceiv-
ably would it have paid attention to the leakage of electric
charges through the air or to the behavior of magnets, wires,
acids, and frogs' legs. Yet from these actually came the
modern lighting devices.

I believe that the same logical and psychological principles operate in the field of government
generally. A highly centralized and organized form of control may be very effective in performing the specific
functions assigned to it in the manner stipulated by head-
quarters, but it is not a favorable type of organization in
which to take advantage of the potential genius inherent in

27

the group which, if given opportunity, may produce better leadership and develop more advantageous objectives and effective operations.

So I believe that experience, logic, and human psychology all support the view that the type of government most likely to be successful in the long run is the one which directs and inspires, but does not too rigidly control; which offers large opportunity for initiative and for criticism; which has faith in the mass judgment of an intelligent group and in the genius which may appear in unexpected quarters—I have faith in the great social invention, democracy.

VI EDUCATION AND DEMOCRACY

It is said in Proverbs, Chapter XXV, second verse, that "It is the glory of God to conceal a thing; but the honour of kings is to search out a matter." . . . In our democracy, the people must accept and share the responsibilities of the government, of the kingship, if you will; and the vigor and growth of their educational enterprises are a measure of their aspiration to the honor of searching out the matter.

It cannot be too often reiterated that in a nation of free men the people must promote untrammeled search for understanding as an essential activity in any program of spiritual and social progress.

As our society becomes more complex by the growth of population and the increasing interdependence of its parts, ever greater wisdom is necessary at every level. Less and less will mere intuition and offhand judgments suffice; more and more will profound knowledge, far-sighted analysis, and skillful operation be required. Wrong decisions will become more catastrophic in their consequences. . . . On the positive side, there are opportunities as never before for constructive leadership in every area of the natural and social sciences, business administration and labor relations, politics, and statesmanship.

Mistakes arising from ignorance, or lack of skill or foresight, may be just as disastrous as actions with subversive intent.

Our educational institutions are faced as never before with opportunities so great that the very fate of mankind will depend on the effectiveness with which they can meet this challenge.

Just as "freedom and democracy must be reborn in and rewon by each generation," so must our educational objectives be re-examined, re-evaluated, and continually re-established on an increasingly sound basis to meet the changing requirements of our evolving civilization.

Even in technological education we are all too prone to give training in the arts of the past instead of the present, and to overlook the fact that we should be preparing our students for the future. . . . The time is opportune to see if our methods and curricula can be more effectively designed to educate men and women for better living and better service in the world, not of the Victorian era . . . but in the world of today and of the next generation.

It is quite generally felt that "knowledge comes, but wisdom lingers," and further that at least one reason why wisdom lingers is that we have lost touch with philosophy amidst the competitive attractions of the ever-increasing number of specialized sciences.

We live in an age of inescapably increasing specialization, and this fact makes a simultaneously increasing co-ordination and co-operation absolutely prerequisite to social progress and even to social stability. Just as "der herr geheimrat" professor of a generation ago is now replaced in our universities by a balanced and co-ordinated group of experts, so also in all phases of our complex life the intelligent planning and administration must come from well-trained co-ordinated groups.

Education has the central role in this matter. It not only provides the experts, but it must also provide training in such a truly cultural atmosphere as will lead to broad mutual sympathies between specialists in diverse fields, and to a realization that true wisdom (or shall we call it philosophy?) exists not in departments, but includes the sum total of knowledge and art and aspiration.

I am told that there used to be an old banner in the Tech on Boylston Street which bore the motto: "To train men liberally to be leaders."

Let us explore the meaning of that word "liberally,"—"To train men *liberally* to be leaders." The word "liberal" is derived from the Latin and means "free." To liberate means to give freedom. A liberal education is one which gives freedom to the mind and spirit of the person thus educated. Elementary physics will illustrate the thought: We speak of the free expansion of a gas. By this we mean that a gas is uninhibited in expanding to fill any container in which it is placed. The freedom of a gas is therefore restricted entirely by the boundaries of the container. If those boundaries be enlarged, the gas expands correspondingly into the new territory thus made available.

One might say that all of us are prisoners within certain confines imposed upon us partly by accident, partly by environment, and partly by the limitations of our own initiative and imagination. The purpose of a liberal education is to push these imprisoning barriers out to much greater distances so that we have the freedom and possibility of thinking and working among the resources of a very greatly expanded world of ideas.

To be a liberal leader a man must have far more than . . . vocational skill. He must have resources on which to call to meet new and unexpected types of problems. If he does not have these resources, he must at least know that they exist, where to find them, and how to make use of them. He must know how to deal effectively with men and with groups of men, as well as with machines and formulas. To be a leader and an effective citizen he must have interests outside his profession, interests which will add to his enthusiasm for living. He must have ideals of right and wrong, of social justice and sportsmanship, which can come to him only through acquaintance with the fine personalities and the noble thoughts that are our heritage of the present and past.

Personalities are the chief concern of men, and therefore of education as well as of social progress. . . . Ideals, ethical principles, values, wisdom, opportunity, satisfaction—these are the worth while things of life, and they are all attributes of personality. Important as science and art may be, and I would challenge any one to defend them with more sincerity and conviction than I am prepared to do, we must look at science and art not as the final objectives, but rather as steps on the path toward the high goal which is a generally satisfactory philosophy of life.

Beyond the borders of our own land . . . we . . . see whole areas of the civilized world in which education, in the sense of a free and orderly development of personality, has been arrested and distorted. In those lands neither the quest of truth nor its dissemination is any longer free; and civil liberty which has been the spearhead of social advance, is blunted into submission to political authority. The name of science has been irreverently invoked for the propagation of gross fallacies through the subjection of the schools. . . . Personality, most precious and potent of all social values, is by way of being stamped out under the heel of regimentation.

Here in our own country, the present picture when fully unveiled is less shocking because less extreme than in some foreign lands; and yet it is disturbing. Our schools, we say, are still free; and yet are they quite free as long as minority groups, from one motive or another, are seeking to impose restrictions upon teaching? . . . Our institutions in general are still free, we say; and yet are they truly and altogether free as long as half-truths and untruths may be broadcast unchallenged to the uttermost limits of the air, while the teacher must think twice before he mentions the soviet, fascist, or socialistic theories of government to his classes, let alone points out any high ideals or good features of such systems? "It makes all the difference in the world," said Whately, "whether we put Truth in the first or in the second place." And whether we in America are to put Truth in the first place, and keep it there, is a question, it seems, not yet definitely answered by our democracy.

One of the great educational problems in our democracy is to build a group or party or national ethics which will come closer than it now does to the ethical standards generally accepted for individuals. Another great problem of education is to disclose the fallacy of meeting danger by burying our heads in the sand like the ostrich, rather than facing situations squarely and honestly.

34

Potent powers in this educational program will be the idealism and individualism of Christianity, the sportsmanship in personal relations which is an Anglo-Saxon heritage and which is certainly growing stronger day by day, and in general the increasing social conscience which manifests itself in industrial relations, legislation, and public opinion as voiced in the press.

Only a part of a really adequate college education is secured in the classroom—a very important part to be sure, but often almost wasted unless accompanied by qualities of character which are largely the products of personal contacts in the activities of community life. Consequently I believe that students in a college should be considered, not as youths studying and reciting under the discipline of a faculty, but rather as young citizens in a community—a community centered around intellectual pursuits but involving the responsibilities incident to the social, recreational, and professional interests of the group.

We are moving more and more in the direction of giving every student, including the best, the opportunity to progress as fast as he can—to keep him up to the level of his best and most ambitious effort.

Much can be said for "equality of opportunity" but I think not enough has been said about the importance of providing exceptional opportunity for students of exceptional talent. In the long run the progress of our nation and the welfare of all are profoundly affected by the outstanding performance of a few. It is important that the "potential few" be found and given every opportunity to develop.

An educational system of complete uniformity, turning out standardized products, would, because of the complex aspects of psychology and modern life, be as unstable as a gas in which all molecules are crowded into the same position: it would explode!

An educational institution justifies its existence and public support in proportion as it renders to society a valuable service; a student justifies the aid which makes his education possible to the extent to which his training is used to serve society (not himself), and he has a moral obligation thus to use his training and the emoluments accruing therefrom for the benefit of society.

I have sometimes thought of a distinction between the great universities with many schools and departments on the one hand, and the individual professional schools, on the other hand, as somewhat analogous to the department store and the specialty shop. In a department store you can find just about everything, and can frequently do so at decidedly bargain prices. Specialty shops, however, devote their attention to a particular line of goods, and the only excuse for their existence is their ability to provide a higher grade of product, fitted more individually to the desires and needs of the customer.

I believe that the problem of maintaining and further improving the *quality* of education in the United States is the most difficult problem facing American education in the years immediately ahead. It is a problem of finance, of management, and of general public appreciation.

VII SCIENCE IN GENERAL EDUCATION

The applications of science so permeate our lives that acquaintance with science and engineering becomes an essential attribute of every educated person. Some of course will specialize much more highly than others, but it is certainly true that no one can consider himself well informed or able to cope intelligently with the problems of modern life unless he has a basic interest in and understanding of those engineering developments which form so important a group of our daily contacts.

I like to think of culture as signifying a sympathetic understanding of life, and in this sense it is evident that the cultured man of the future will be one who includes among his interests and attainments a considerable knowledge of science and engineering.

It is important for the welfare of the country that our colleges prepare young men and women with a training in science which will make them competent to contribute effectively to the solution of these problems. It is equally important that the rank and file of our population, and especially our future political leaders, be given a sufficient understanding of science to enable them to appreciate what science can do and what conditions are prerequisite to its effective operation.

It is not possible to understand our environment or to live intelligently in it without an understanding of science and engineering.

The mental discipline inculcated by the study of science and engineering is, I believe, at least as important an aspect of their cultural value as is the acquirement of knowledge regarding our surroundings.

Science requires straight and independent thinking. Every hypothesis or idea is capable of definite proof or disproof. The habit of mind that subjects every idea to rigid test is of utmost value. Much of the loose thinking in social, educational, political, and economic affairs would be avoided if the workers in these fields could be given a real training in accurate scientific thinking.

Engineering training, besides developing habits of scientific thinking, gives the additional valuable habit of accomplishment, the power to marshal all available resources and arrive at the desired result.

Scientific and engineering studies possess the same advantages as do the classics in giving a training in precise thought and expression, but in addition they afford training in observation, in inductive and deductive reasoning, in exact checking of hypotheses and assumptions.

I well remember my first class in geometry. The professor took a piece of chalk in each hand, went to the blackboard, started to draw two converging lines and, as he did so, asked a young lady in the class: "Miss Harris, are these lines going to meet?" "Yes, sir," she replied. Whereupon he turned them away from each other and finished them as nonintersecting, divergent lines. "Why did you say they were going to meet?" asked the professor. "They looked as if they would," replied Miss Harris—at which the professor closed the incident with the remark: "Now Miss Harris, if everything in this world depended on looks, where would you and I be?"

There is nothing more important in our present education than to develop the habit of looking for the facts and fundamental principles, and training ourselves in the habit of making decisions in accordance with the facts rather than in accordance with hunches, preconceived ideas, prejudices, or a general policy of blind optimism as to results. It is this type of accurate and impersonal approach which has made possible the extraordinary development of science and engineering.

A man whose mind is trained to view situations objectively, to draw rational conclusions from observed facts, to plan his course intelligently in the light of these facts and conclusions, is a man who is a safe citizen in a self-regulating society, i.e., in a democracy. On the other hand, a man who is not trained or capable of thinking rationally, who is governed by his emotions and prejudices, is an unsafe member of a democratic society,—an element of instability.

Once I was asked: "Why do you think that students should study science even if they do not expect to become professional scientists?" I am afraid that my answer was somewhat flippant when I replied: "If your students plan to go through life as human vegetables or contented cows, I see no reason for their study of science."

But this answer was not altogether flippant. Study of science means inquiry regarding the nature of the world in which we live. It implies curiosity regarding the nature of the materials of which the earth is composed and the forces which make these materials behave as they do. It implies curiosity as to why and how things grow, why people behave as they do, what causes sickness and how can it be prevented and cured. It implies a desire to use this knowledge to some purpose, whether it be to make something or to do something that we think would be of some benefit to us.

As I see it, it is this feature of purposeful curiosity, and ability to exercise it, that most differentiates us from the vegetable or the contented cow. Speaking broadly, therefore, I would say that the scientific spirit is one of the chief characteristics of human intelligence, and of course science should form a part of the interest and education of every person.

Science has a valuable part in education because it creates knowledge, disciplines the mind, and has great utility. You may study it from a sense of duty, because of these three values, or you may study it for fun. I suspect the latter is the real motive power and the former is the excuse behind most real scientists.

However it is viewed, science seems destined to occupy an increasingly significant place in our educational system.

VIII TEACHING SCIENCE

Undoubtedly, the easiest and laziest method of conducting a class in science is to assign a lesson from a textbook, quiz the pupils on their mastery of this lesson, and then assign the next lesson. In more advanced university work, an equally easy method is to lecture. Lecturing satisfies the self-esteem of the lecturer, who takes artistic pleasure in the logic and skill with which he covers his subject, and who avoids nearly·all contacts with his students which might disturb the perfection of the presentation. Undoubtedly, also, these are the world's worst methods of teaching science.

A real science teacher is far more than a task-master or a stoker, and science does not consist of learning lessons by heart or taking notes. In fact, an argument can be made that the habit of learning by memory, with which so much of our education is concerned, is a handicap rather than an asset to real mastery of the method and spirit of science.

Let us consider the question of lectures from the standpoint of training a research physicist. Is the best lecturer one who explains so lucidly that you are left with a comfortable feeling of understanding, or one who, out of the fullness of his experience, opens out before you the un-attainable yet challenging expanse of the subject?

It is said of the great Agassiz that his first assignment was to give his students a specimen of something, a fish or a flower, and leave them for three days with absolutely no instructions except to observe.

Science is not a technique or a body of know-ledge, though it uses both. It is rather an attitude of inquiry, of observation and reasoning, with respect to the world. It can be developed, not by memorizing facts or juggling formulas to get an answer, but only by actual practice of scientific observation and reasoning. The teacher, to be effective, must have the same attitude as the pupil, after the good old method of Socrates.

There is no stimulus like the joy of discovery, and it is often a wonder to me that any interest of students in science ever survives the year upon year of learning to which they are often subjected without ever tasting the joy of an original discovery or idea, however elementary.

There is all the difference in the world be-tween running a laboratory to verify the laws that have been learned in the textbook, and running it to bring out or suggest these laws in advance of the textbook. The latter method is slower, and far more difficult for teacher and

pupil alike; but the former method is not really science at
all, merely illustration and technique.

The best and also the easiest teaching is
done by the professor who is working with his graduate
student on a research problem. Here all thought of ped-
agogy is thrown to the winds; the teacher and student are
collaborators on a job that taxes the resources of both. The
student learns by example and by his own mental effort.

I firmly believe that no element in the training and environment of science teachers is so valuable as
that which keeps them in continual live touch with the
progress of science itself so that, through their own interests
and example, their students may see science as a live subject
and feel that they share some part, however insignificant,
in its progress.

Success in the teaching or the study of science
is in proportion to the extent to which the spirit of inquiry
becomes the motive power.

The qualifications of a research physicist . . .
consist of aptitudes and attitudes. . . . The first qualifica-
tion of a research physicist . . is that he must be born
that way.

The essential aptitudes are first, a space-
sense that will enable him to devise tests, and second, a

retiring personality that gets pleasure from quiet accomplishment without the need of public applause. The research
worker should in addition, be endowed with an analytical
ability and preferably but not necessarily, manual dexterity.
To these we may add the healthy qualities of energy and
enthusiasm, without which all success is circumscribed.

How do these considerations and facts affect
the question of the training of physicists? In two ways. We
cannot inculcate or cultivate these traits. We can only add
to them certain attitudes and habits, and a small store of
facts. Without the inborn aptitudes, our trained physicist
is like an overrigged ship, destined to founder.

Not only must the research man have the
necessary aptitudes, but he must have no extra ones.
Unused aptitudes produce dissatisfaction and half-
heartedness.

Assuming that we have chosen our men,
what can we teach them? Not very much. The requirements of a research man are different from those of an
engineer. . . . I will quote the opinion of one of the greatest
research leaders, Dr. W. R. Whitney: "The asset of engineering is exact knowledge. The valuable attributes of
research men are conscious ignorance and active curiosity."

Are we teaching or can we teach these quali-
ties? We have schools reputed to be so learned that you can

always tell their graduates, but can't tell them much. Have we any that teach conscious ignorance?

The value of this attribute may not at first sight be apparent. Apart from the fact that the know-it-all graduate is unbearable to others, is unwilling to stoop to the tasks for which he is fitted, and is largely immune to further learning, there is a very special significance to research workers of this attitude of conscious ignorance. *If you know the answer beforehand you will always find that your experiments yield that answer.*

Conscious ignorance is a negative quality and becomes dynamic only when joined to active curiosity. . . . It may be more than a myth that the elephant who couldn't restrain his curiosity and wears an elongated nose in token of it, developed into one of our most intelligent of animals, while the ichthyosaurus is extinct.

It is remarkable how little most people see. . . . Not only do we not observe, but we don't question. Ask the average physicist why water rises in a small tube and he will answer: "Because of capillarity." Has he added any information in substituting a long word for your short one? Names are not knowledge. If our curiosity is of the kind that is satisfied by a name, it will not lead us far.

Active curiosity, curiosity that not only craves to know and is unsatisfied by names or authority but does something about it, is the most valuable attribute of a research man.

Conscious ignorance and active curiosity are attitudes which may be caught like contagions, cultivated like choice flowers, and retained to old age.

Consciousness of ignorance combined with active curiosity, is the spirit of youth. Beware of the man who tells you, "I did that twenty years ago." He is through, either as investigator or as a useful teacher.

IX ENGINEERING EDUCATION

The question of engineering should be of interest not only to those of us who are engineers, but to the entire public which lives in an engineering world.

Engineering education is the *sine qua non* of this technical age. Unless it is effective and adequate our type of civilization cannot go forward. To be effective, it must be progressive, for engineering art is not static; it is very dynamic.

For the benefit of society, as well as for the most efficient work of the engineer, it is essential that the engineer should be trained to think not only of his specific engineering projects, but also of their larger significance in the economic and social order.

An English wit lately observed that since women have found they could talk about everything, they won't talk about anything else. I sometimes feel that now that engineers have begun to discuss the social aspects of engineering, we too should refuse to talk about anything else.

Lest this be taken as something more than a soft impeachment of an oratorical fashion, I hasten to emphasize that I am proud of this new voice of engineering, whether oratorical or not; proud of this growing insistence on a proper recognition of society's indebtedness to the engineer, and the engineer's widening obligation to society. Out of this chorus of reiteration there must emerge a wider recognition among our own votaries that if engineering is to be a true profession it cannot be restricted to a plane surface of technical proficiency, but must embrace the third dimension of social responsibility and awareness.

To provide a course of study which will serve as a sound foundation for scientific or engineering work and at the same time give adequate consideration to the relationship of technical developments to the orderly progress and growth of our social and economic structure, is not only an opportunity but actually an obligation upon those institutions which are equipped to do so.

The time is passing if not already past, when the bright young boy with ingenuity can make effective contributions to the improvement of engineering or of industry. To his brightness and ingenuity must be added a training which is broad and thorough in its fundamentals, and yet specialized in its ultimate direction. Our educational institutions are our most effective means of providing this training.

We cannot get far by trying to impose an engineering education, however excellent it may be, on a young man of mediocre ability or one temperamentally unfitted for technical or administrative work. The idea reminds me of an experience which my sister had . . . in India. She had engaged a native electrician to install some new fixtures in her house, but he seemed particularly stupid and kept coming to her for instructions. Finally, in exasperation she said to him: "Why do you come asking questions all the time? Why don't you use your common sense?" "Madam," he replied gravely, "common sense is a rare gift of God. I have only a technical education."

There are two principal reasons for increasing training in scientific principles and the more fundamental aspects of engineering, as opposed to going farther into the field of specialization in the training of engineers. The first of these reasons is to be found in the development of the engineering specialties themselves, starting first with simply applied science followed next by the branching off into the various main engineering fields, such as electrical, mechanical, civil, etc., until with further discoveries and developments there arose subdivisions of these fields, such as radio, hydraulic, refrigeration engineering, etc. Now, even these subdivisions are becoming so highly specialized as to indicate the need of still further subdivision in the direction of specialization.

The situation confronting the educational institution is therefore this: the limitations of time and

51

mental capacity make it utterly impossible to train a student in all of these specialties, and if the attempt is made to train him even in a few specialties, the result is a narrow training and one which lacks the vigor which comes with a clear grasp of fundamental scientific principles.

It appears, therefore, that the best training for effective service must come with the specialties developed only so far as they can be carried without losing the opportunity of training in the underlying principles. The more highly specialized activities can then be learned either in postgraduate work or in actual contact with the practical problems in the work of the engineer. . . . At the present time large industries are equipped to train their own men in the various techniques of their work and prefer to do so.

The technique of various engineering processes and these processes themselves, are frequently subject to change. Consequently the training in a technique may not be of lasting value, whereas the training in scientific principles is valid whatever be the modifications of their applications.

I do not mean to question the need for accurate detailed knowledge of the job which one is doing. . . . But there are two ways of knowing facts: one may know them by sheer brute force of memory with tremendous exertion or one may know them easily and adequately by seeing them hang together as the illustrations of a body of principles.

52

The man who knows principles gathers facts easily because facts have meaning for him. Principles are labor-saving devices.

Further improvements in engineering education may be expected to follow along a path which was laid out by the chemical engineers. . . . Just about a generation ago a very important improvement was made in chemical engineering education. Attention was focused on the many various typical *operations* which were common to many types of chemical manufacture—operations such as evaporation, distillation, transfer of heat, mixing, grinding, flow of liquids through pipes, etc. By studying thoroughly the basic theory of such processes and the techniques for handling them, . . . the students develop a *power of handling situations* rather than absorbing a mass of specific detail applicable only to appropriate specific situations.

More recently in the development of a program of biological engineering, based upon physical, chemical, and biological operations, a similar attempt has been made to synthesize an appropriate training for the handling of a great variety of biological situations, whether they be in the food industry or in the hospital or medical or biological research fields. I suspect that there may be other directions in which an analogous approach may be made to simplify the educational program and at the same time increase the power acquired by the student.

With increasing specialization comes also an ever more important need for integration between specialties. . . . The first requirement for integration is to realize the existence of the problem; the next step is to take various obvious and informal measures for remedying the situation. But experience has shown that these are not enough, and some policies of organized attack on the problem must be adopted.

Of very great advantage, in my judgment, is a physical layout whereby the buildings which house the various specialized departments are all interconnected under one roof. Since the war we at M.I.T. have taken another step which is already showing excellent results, which is to "federalize" certain new laboratories. Let me give an illustration.

The subject of electronics, both in its scientific development and in its practical applications, is of about equal interest to the departments of physics and electrical engineering, and of somewhat lesser interest to a number of other departments. Some of the work being done by our electrical engineering staff and students might just as well be carried out in the physics department, and vice versa. Similar facilities and in many cases similar problems are involved.

Rather than establish separate and duplicating facilities for the two departments and maintain a compartmentalized approach to the problem, we have established a "federalized" electronics laboratory to serve the common needs of these two departments and of any other department which has an interest in this field and a need for laboratory facilities.

We are convinced that this scheme of "federalized" laboratories . . . is the successful answer to the problem of departmental integration. We shall certainly find other opportunities in the future for similar arrangements, and I anticipate that the logic and success of this plan may lead to its widespread adoption in other institutions.

Of even more importance than specialization and integration in the curricula and laboratories of the various professional fields, is the development of students to be broad-gauge citizens of the world as well as professionally competent scientists and engineers. . . . How to achieve this goal is a problem for all educators but especially for those concerned with professional education, where the requirements for professional competence are so demanding as to leave but limited time for the other aspects of general education which make for "the good life."

Everyone would agree that of first importance in achieving this objective is the having on the faculty men who themselves exemplify this ideal and who have the interest and personality through which it may be imparted to their students.

55

In addition to the influence which comes from the personality and the example of teachers, much can be done by suitable organization. For example, the students' extra-curricular activities offer a rich opportunity for the development of a sense of responsibility, a technique of management, and a spirit of team play. They can also provide opportunities for the stimulation of cultural interests of wide variety.

Still another contribution to this objective is made through that part of the curriculum which falls under the somewhat inadequate heading of the humanities.

What is the purpose of these courses in humanities? Is it to make our students more literate? Is it to enrich their lives by acquainting them with the cultural heritage from the past? Is it to develop ability to get along more effectively and agreeably with their fellows? Is it to give some insight into the current problems of the world today? Its purpose is all of these and more. I do not know how to define it in a few words. The term "humanities" rather well covers the general idea if one takes its original Greek meaning, but in many quarters it has come to have a specialized use which is narrower than we have here in mind. The phrase "general education" has some merit, but is too vague and hackneyed to be very attractive for our purpose. We have not found the right name.

In some respects it is advantageous not to have a name, because we do not want to differentiate this aspect of our educational program too strongly from the professional aspect. They are really all parts of a unified program aimed at developing a man who will be a competent operator in some field of specialization and who will at the same time have the insight, appreciation, and viewpoint which will enable him to find interests, to operate effectively, and to live with satisfaction in whatever community or situation he may find himself. This is a very large order. Obviously we can not achieve it wholly.

If the engineer is to bring his influence to bear on broad public questions he must approach them, not with technical arrogance, but with sympathetic understanding. If he is to counsel the people, he must gain the confidence of the people, and this confidence is obtained by placing ministry to the public above all other considerations. This concept of ministering to the public welfare, which is the concept underlying the professional attitude, is the remaining principle that needs to be fully synthesized with other elements that have been combined to form the engineering philosophy. If we are to be a true profession, we must embrace this third dimension of social responsibility and public service.

An engineering training is an excellent training for a successful administrative career. The

type of young man to whom engineering appeals as a
profession is often the type naturally well fitted for
administrative success. Men of this type like to figure
out ways of doing things and, when a satisfactory plan
is found, cannot restrain themselves from putting the
plan into operation.

This situation would seem to indicate the
advantages to be gained by including more study of
economics and business principles in the curriculum
of engineering schools, not with the thought that such
courses *make* great business men, but rather that a
background of economic principles and knowledge of
business practices . . . should strengthen the engineer
for handling the administrative duties which experience
shows are likely to come his way.

The important thing to remember is that
"the engineer who accepts responsibilities of management,
must put to practical use the products of the social as well
as the physical sciences," and should profit by training
in both.

The spirit of science should be emphasized
in the training of engineers. . . . If every engineering
student could, either by his own investigations or otherwise by close contact with some important field of research,
be given opportunity to develop the enthusiasm and
imagination which go with discovery, then I believe that

a very important element would have been added to his training.

Science and research should play an even more important role in the programs of institutes of technology than they have in the past, and this for educational reasons. The reason is plain: new products, new industrial arts, new ideas are being developed at increasingly rapid tempo. . . . Therefore students of technology in our institutions must be educated more than ever for the opportunities of the future rather than for the techniques of the past. This can be done only if the institutions create within themselves an atmosphere and environment of technological progress through research; if staff and students alike are participating in this progress.

X RESEARCH AND PROGRESS

Many examples might be quoted to prove that pure scientific research, carried on simply with the objective of discovering new truth, is the best long-term investment which can be made, since it is by such work that new facts and principles are discovered, on which the inventive genius may construct new industries and open up new possibilities of human achievement.

To the pessimist I would say that experience ... proves that scientific research is an essential safeguard for your investments; it is necessary to prevent your competitor's getting the best of you in your business. To the optimist I would say that scientific research is the most intelligent gamble—or the best investment for the future—that can be made, and that it is through the products of such research that the direct improvement in the physical condition of men, and indirectly their opportunities for better intellectual and social conditions, must be achieved.

Research is like a snowball rolling downhill. It gathers mass and momentum at a constantly accelerated rate.

The basic reason for the rapid increase in the rate at which science is usefully applied in the various branches of engineering, is that the probability of finding a solution for some practical problem increases very rapidly with the number of scientific facts which we know.

For example, if there is one practical problem which I want to solve, and just one scientific fact which I know, the chances that this scientific fact will solve that practical problem are very small. If, however, I know two facts, I have a much better chance since either fact alone may solve the problem or a combination of the two facts may be the solution. Similarly, if I know ten facts, I have probably a hundred times as many chances to solve my practical problem as if I know only one fact. . . . Perhaps we can say that the rate of development of engineering is proportional to the square of the amount of our scientific knowledge.

While there is much truth in the statement that necessity is the mother of invention, it has often been pointed out that it is far from true that necessity is the mother of discovery. Discoveries come often most unexpectedly in the pursuit of knowledge by the curious and observant. The great background of natural phenomena which have thus been discovered form an immense reservoir from which may be drawn natural laws or combinations of phenomena which can be made to work for the solution of men's needs or desires when necessity arises.

I think the story of research development is well described by the ditty:

> "Little drops of water,
> Little grains of sand,
> Make the mighty ocean
> And the pleasant land."

Anyone familiar with scientific literature realizes the enormous number of contributions, most of them small and not very significant, but each and all gradually raising the level of understanding in the storehouse of knowledge until finally the stage is reached at which a great scientific discovery or a mighty practical application can be made.

The securing of practical results from research was described by Kettering as being like the scattering of seeds in nature. Only a small portion of all the seeds scattered from the trees and the plants mature. Most of them are failures, but it is those which do mature that maintain life on this earth. A sizable proportion of scientific research work turns out to be a failure in the sense that it leads to nothing significant. It must always be so, however, because one fact will always be true in scientific work, namely, that no discovery of any great importance is likely to come out of any research whose end result can be foreseen at the beginning.

Following Kettering's analogy we may specify certain requirements which must be met if a good research crop is to be secured. In the first place, many seeds must be planted. In other words, many ideas and experiments are called for. In the second place, the soil must be tilled and fertilized. In other words, the work needs encouragement, facilities, support. In the third place, time must be given for the work to mature. Sometimes this is short, sometimes long. In the fourth place, the harvester must go in at the right time, not prematurely. In the fifth place, skilled planters, tillers, and harvesters are a prerequisite. No plant in nature is more tender or requires more careful nurture than does a research idea.

The war experience demonstrated that the so-called pure scientists and the so-called applied scientists are in general not two different breeds, but are really brothers under the skin, for top-notch pure scientists turned out in very many cases also to be top-notch applied scientists, and the reverse was also true.

The experiences during the war, as well as the growing experience with the great industrial research laboratories and of some of the groups in university laboratories, have increasingly demonstrated the great power of co-ordinated group effort, where a number of scientists or engineers with different backgrounds and points of view can tackle a major scientific problem in a co-ordinated manner.

The most important new development in our universities is the emergence of the recognized "research program" as differentiated from the multiplicity of more or less unrelated "research projects." These research programs are large projects built around very important objectives and generally involving the co-operative effort of men from various departments and with various specialties. They involve some sort of special organization within the university which is different from the ordinary department because it is not particularly involved with curriculum or degrees, although it may contribute to both. It generally involves very substantial financial support which usually has to be secured from sources outside of the regular funds available to the institution.

There is another healthy aspect of this situation, namely that these organized groups in various institutions are showing a most desirable tendency for inter-institutional co-operation, so that this program within the institutions is actually in many cases a national program involving the institutions, industry, and government. In other words, it involves the members of that great team which was so effective in our national war effort.

Whenever a great new research laboratory . . . is established, it should be a source of satisfaction and gratification to every citizen because in a very real sense it is an addition to his wealth: its existence insures for him a greater degree of security and a higher standard of living than he could otherwise enjoy.

Perhaps the greatest need and a fundamental one is to select and train and get into the right environment and on the right jobs the young men of real genius and ability. There is certainly no way in which money can be wasted more rapidly than by turning research over to people of pedestrian qualifications.

It is men and ideas which have constructed our past and which will forge our future.

With free schools and untrammeled—nay, an ever more vigorously prosecuted—science, and, finally, well developed philosophy to steer our life, social progress toward the worth while ends of living is assured. It is the business of education to keep bright the torch which lights our upward and difficult path, and our business as educators to . . . discover ever brighter torches and more direct paths.

A couple of years ago a small group, representing the professional fields of theology, law, medicine, business, and engineering, met to exchange ideas on the subject of professional education. In one of the sessions, a strong case was made by a biologist for the thesis that the basic motive back of education and in fact, back of everything else in life, is the "urge to live." . . . Truly the urge to live is basic; it is a feature of biological behavior of cells and tissues, of the instinctive reactions of all animals, and of the conscious planning and effort of human beings. But though mankind shares with all other forms of life the urge to live, there seems to me to be another urge which is peculiar to man and distinguishes him from all other animals or other living things. It is the urge to live *better*.

It is man alone, of all living things, who has consciously and to a staggering degree, changed his ways of living. . . . Man has employed imagination and logic, he has invented, he has developed new skills, he has created new concepts of values, he has manipulated the materials and forces of Nature for his purpose. He has done all these things because of his urge to live better.

While it is occasionally pleasant to think back, it is far more profitable and interesting to think ahead. Adventure and progress and exhilaration of achievement always lie in the future, and their planning should be the chief concern of the present.

This book
has been composed in Photon Basker-
ville, printed, and bound by the M.I.T.
Printing Service to designs by Marilyn
Fraser '55, for the Undergraduate
Association, under the direction of
David L. Rados '55, Roy M. Salzman
'55, and R. Peter Toohy '55. The
frontispiece is from a photograph by
Robert H. Tucker '56.

www.ingramcontent.com/pod-product-compliance
Lightning Source LLC
Chambersburg PA
CBHW031814190326
41518CB00006B/333